The Interplanetary Laser Jousting Tournament

or

Relativity Revised

by John Barry

DORRANCE
PUBLISHING CO
EST. 1920
PITTSBURGH, PENNSYLVANIA 15238

Dorrance Publishing Co
585 Alpha Drive
Pittsburgh, PA 15238
Visit our website at *www.dorrancebookstore.com*

ISBN: 978-1-4809-9518-5
eISBN: 978-1-4809-9590-1

Table of Contents

Chapter One: Does Light Travel at Different Speeds?. 1

Chapter Two: The Tournament. 9

Chapter Three: But What Would Really Happen?. 13

Chapter Four: More on Optical Illusion Relativity and Different Speeds of Light. 23

Chapter Five: Not Only Would Time Not Slow Down, but Space Wouldn't Contract Either 27

Chapter Six: Quantifying Relativity. 31

Chapter Seven: Relative to This or That?. 43

Chapter Eight: Super Lucent Speed 59

Chapter Nine: An Actual Example of Relativity 65

Chapter Ten: Why Are There No Examples of Things Going Faster Than Light?. 71

Chapter Eleven: Another Experiment. 75

Ilustrations_

A Fast and Slow Light Separator 6

A Fast Light Detector. 7

A Bell Curve Showing the Amount of Light Going Faster or Slower Than the Given Speed of Light. 20

A Bell Curve Showing the Amount of Laser Light Going Faster or Slower Than the Given Speed of Light 21

A Graph Showing Real and Apparent Speeds of Objects Coming and Going 37

A Graph Showing Real and Apparent Lengths of Objects Coming and Going. 39

A Graph Showing How Angle of View Affects the Appearance of Relativity. 41

A Diagram of the Apparatus Used in the Michelson Morley Experiment. 45

A Diagram Showing the Sagnac Interferometer Device 63

Related Reading

Einstein's Brainchild Relativity Made Relatively Easy, Barry Parker.

The Dreams that Stuff is Made Of: The Most Astounding Papers of Quantum Physics and How They Shook the Scientific World, Stephen Hawking

One, Two, Three... Infinity, George Gamow

Preface

As all the easy scientific discoveries are made, as the basic facts and principles are understood, that which remains is a little more esoteric. The best way to advance scientific discovery is to test hypotheses by performing experiments and observing natural phenomena to build theories that are correct – that stand up to newly discovered realities, to the test of time. But easy scientific discoveries become rarer. Experiments become more challenging and more expensive. Natural phenomena that display new truths are more obscure and better hidden.

Aristotle believed that the careful use of good logic was the proper way to understand nature. Lavoisier demonstrated that by quantifying your observations and using mathematics, to analyze the results that you could push what you learned from experimentation a lot further.

But we can't get away from those difficult and costly experiments to prove the reality of what "has been established to be true."

Juvenal once wrote, "Everybody is writing satires nowadays, so why shouldn't I do the same?" I'm afraid that the only other excuse that I have for putting an unproven hypothesis in a book is that mine appears more logical than the other hypothesis that is offered in the Theory of Special Relativity and that I suggest a few experiments to test my hypothesis.

It will be a wonderful gift to the world when someone performs the experiments that demonstrate the truths of light and special relativity. After all, some day we will want to travel through space at very high speeds.

John Barry

Chapter One

Does Light Travel at Different Speeds?

The stuff of light, photons and waves of electromagnetic energy, comes out of candles, stars, lightbulbs, and radio towers (not light but electromagnetic energy in longer wavelengths) at the speed of light. Is it that it comes out at no other speed? Why shouldn't there be more, lots more, of these quanta of energy/infra-sub-atomic particles, some traveling slower than, others traveling faster than the speed of light? The speed of light is not a barrier. Anything can travel faster than the speed of light.

Before we get to our story, let's consider an imaginary experiment. It's imaginary because it's a little too challenging for me to conduct. It's the sort of experiment that our scientists may some-day perform on the International Space Station.

It goes like this: Get a source of very bright, white light and aim a tightly focused beam of the light at a rapidly twirling mirror. This will be similar to what Jean Foucault did in 1849, but he only wanted to measure the speed of light. We are doing something different. Around the twirling mirror in our experiment is a black curtain that absorbs the light reflected by that mirror. The light source is placed high enough that the light from it goes over the curtain, but any that misses the mirror is absorbed in the curtain behind the mirror. There are two small gaps in the curtain to the right of the light source. (I have the mirror twirling counterclockwise when looking down on it.) When light reflected from the twirling mirror passes through the first hole, it hits a stationary mirror and is reflected further to the right to a second stationary mirror that reflects the light through the second small gap in the curtain back to the twirling mirror. The first gap is, from left to right, only a little wider than the beam of light, and there are black curtains behind the stationary mirrors to catch any stray light. This experiment does have something to do with the speed of light, so the two stationary mirrors are to be placed far enough away from the twirling mirror and far apart from each other to give the light enough room to do its thing. The light that comes back to the twirling mirror is reflected again over the curtain toward a distant periscope that is turned sideways. The light goes into the opening of the periscope, is reflected to the left by the first mirror, hits the second mirror, and goes out of the periscope. The second mirror is stationary; the first mirror can be adjusted to the right of where the beam of light hits it. The periscope is mounted on a large black card covering a small window to an even darker chamber. There is a rectangular hole in the card to accommodate the aperture as it moves from left to right. The moving part of the periscope, itself, is mounted to a smaller black card that covers the first card's hole whatever position the periscope is in except for, of course, the moving aperture which is the only thing that is exposed. We don't want that light to get through, except through the lens of the periscope. After going through the periscope, the light hits the edge of a very large wheel, let's say a mile in circumference, at a tangent. The wheel is notched all the way around it. One side of each notch follows a line of radius, it is in line with the center of the wheel but is only an inch deep. The other side of the notch is at a right angle to the first side and a few meters long. Just long enough to reach the edge of this huge wheel where the next notch starts. Placed in the short side of each notch is a mirror, one inch wide; it fills the notch. The length of the notch and the mirror that is glued into it are enough so that, as this very accurately built wheel spins, the beam of light never misses the next mirror. (It could be that it would be better to just mount the mirrors on the surface of the wheel, but I whimsically prefer this way.)

Part of the way around the wheel and not too far away from it is a curved mirror facing the wheel. This huge mirror is meant to reflect light back to the spinning wheel almost exactly to the same point that it came from but just a tiny bit toward the center so as to hit a wheel mirror that is to the right and closer. This time, I have the wheel spinning clockwise, and the beam of light first hitting it on the left side. The wheel mirrors are always backing away from the light because the mirrors are receding from the light as they spin around the wheel. The huge curved mirror's curve would be shaped so that its surface would be progressively perpendicular to the paths the beam of light will follow along lines of tangency to the wheel. (The shape, I suppose, would be a portion of a circle or a spiral.)

The entry side of the curved mirror, where light first going to the spinning mirror, passes it to the left looking at it from the periscope, is farther away from the wheel than the exit end of the curved mirror. The beam of light coming out of the periscope is aimed to hit the wheel tangentially, but we aim the beam slightly off tangent, more toward the center of the wheel so that it lands squarely on the first mirror, bounces to the curved mirror, and then back to another notch mirror a little farther up the wheel. The side of the notches that the mirrors rest on may have to be trimmed a tiny bit greater than 90 degrees to the long side of the notch, that is to say, slightly off the radius line, for this to work. We don't want the beam of light to be absorbed by the long side of the notches. Coming from the periscope, the beam of light just misses the edge of the curved mirror. When it is reflected back from one of those mirrors on the wheel, it lands squarely on the edge of the curved mirror. The surface of the curved mirror is close to perpendicular to a tangent line but actually parallel to each notch mirror, and since the light first struck a spinning mirror slightly off tangent to the wheel, it will always strike the curved mirror off perpendicular and return to the wheel and strike another mirror in a position slightly further around the wheel to the right.

For this to work, the curved mirror is turned so that the away end of it is closer to the wheel and light bouncing from a point on it at the edge of the far-away exit end strikes a notch mirror closer to it but in the same near-perpendicular way and is reflected back to narrowly miss the edge of the curved mirror. The light beam climbs a ladder of notch mirrors to get from one end of the curved mirror to the other. Now having reached the other end, that beam of light that has narrowly missed the curved mirror goes to a device that measures its intensity. (A very sensitive device because by now the light must be pretty attenuated.)

When the wheel is not spinning, the light passes through this Rube Goldberg contraption, and as we slowly adjust the periscope opening to the right, the amount of light that our photometer measures is reduced steadily to none at all. Now let's reset the periscope aperture back to the left so the visible beam of light enters it and try it again. This time, we will spin that huge wheel extremely fast. Let's say 6000 rpm or 100 revolutions per second. (You see why this experiment is one for the weightlessness and vacuum of space.) Each notch-mirror is traveling away from the beam of light at 100 miles per second. Multiply that times the number of times that happens going through the curved mirror, let's say 100 times, and the light has been slowed down by 10,000 miles per second before it reaches the photometer. That certainly should be a large enough fraction of 186,000 mps, the speed of light in space, for the change in speed to be detectable. For one thing, there would probably be a noticeable red shift in the light that hit the meter.

Now, if Albert Einstein's Special Theory of Relativity is right and light travels at only one speed, the given speed of light, then spaces between the wheel's edge and the curved mirror contract, and time slows down so that that light can hit the meter at the speed of light. The photometer continues to detect light and moving the periscope's aperture to the right darkens the input to the meter. But what if light travels at any speed but we can only see (or detect in any way at all) light and other electromagnetic

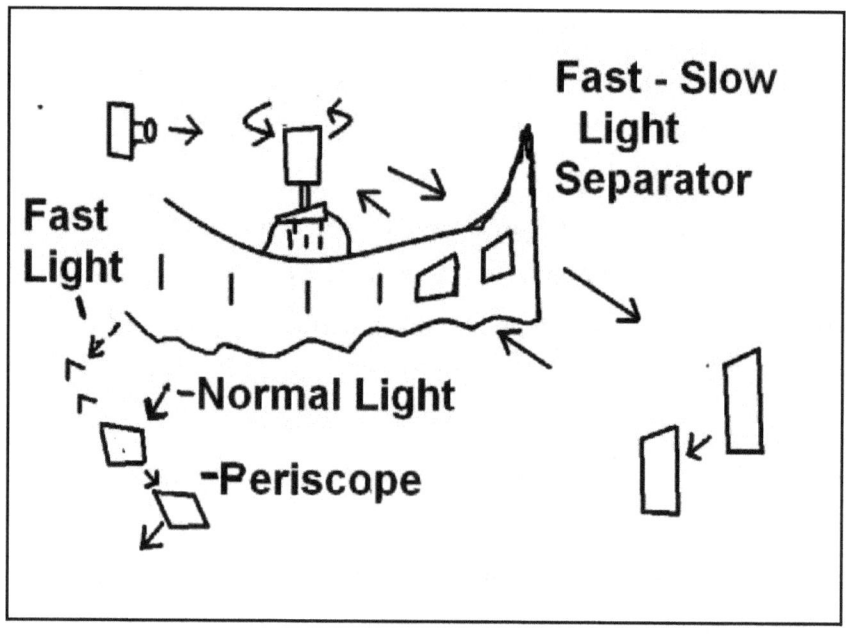

A Fast and Slow Light Separator

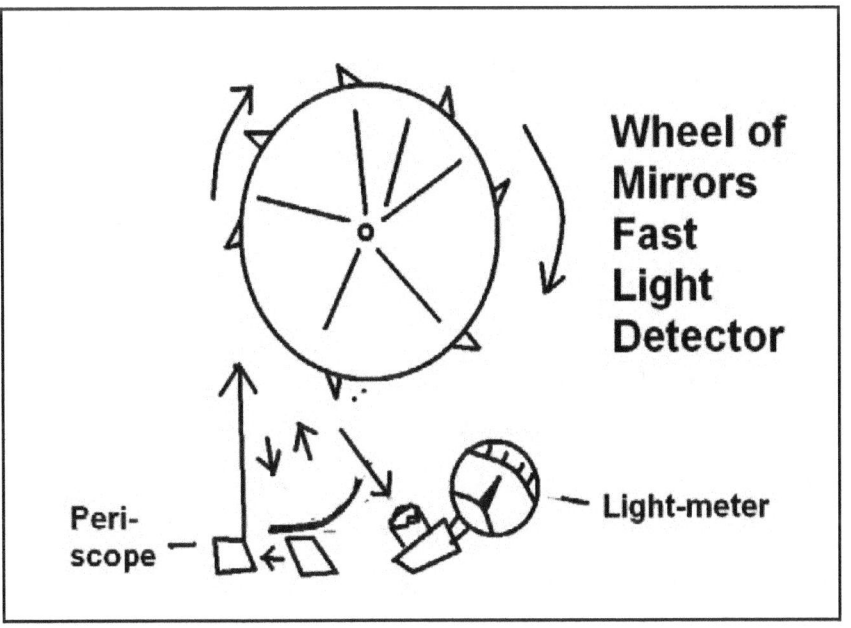

Wheels of Mirrors Fast Light Detector

energy that comes to us at the given speed of light? Then light that struck the rapidly twirling mirror and went through the narrow gaps in the curtain, is composed of slow undetectable light that left the light source a tiny bit early and regular light and fast light that left a tiny bit late. We only see the normal light, but all three kinds of light struck the twirling mirror at the same time to head for the first stationary mirror. On their journey to and between the mirrors and back to the twirler, they will get spread out a little. The fast light re-hits the twirler before the rest and is bounced a little to the right of the periscope aperture; the normal light hits the aperture, and the slow light, we won't worry about. Now when the visible light goes through the spinning wheel and curved mirror apparatus, it gets slowed down to below the speed of light. It is now undetectable, and the meter reads nothing. But as we slowly adjust the periscope aperture to the right, the volume of light measured by the photometer increases. We are seeing the faster light that is being slowed down to the regular speed of light. If, as the special theory has it, space does shrink, and time does slow down, and it is the originally visible light that is hitting the photometer then that light will be red shifted because the contraction of space or slowing of time stretched the light. But if this is different light, faster light, then the light is red shifted just the same, but only because it came out of its source at the same beat (that is to say, the same frequency or spectrometer signature) as normal light but going faster. Its frequency was the same as normal light, but its wavelength was longer from its very beginning inside the light source. If only someone could perform this experiment, so we could find out if this is so!

Chapter Two

The Tournament

And now we embark on a journey to a galaxy far, far away. Er, no, not that galaxy—some other galaxy somewhere else.

The peoples of two planets who were rivals were intensely competitive. They were both very technologically advanced, so they decided to hold an interplanetary laser jousting tournament to see which of them possessed the greater military prowess without actually going to war. They chose an empty place in the inky blackness of deep space between the two planets and positioned stationary camera satellites along a 186,000-mile-long jousting course (which, other than the brightly lit and motionless satellites, was just empty space). There was a satellite at the midpoint with cameras facing both ways and toward the middle to catch the action as the two spaceships passed each other and a satellite with cameras at either end to mark each spaceship's entry and exit gate. (Nothing more than a point relative to the satellite.) Although the spacecrafts carried crews, they were controlled by very advanced and precise computers so that they would both enter the course at the same fraction of a second precisely at the gate where they were supposed to enter, fire their mighty laser cannons when and where they would have the best chance of hitting the other craft, pass the other craft on the proper side of the midpoint camera satellite, then fire their rear-pointing laser cannons at the opponent, and then exit the course through the proper gate. They would pass very close to each other and be going very fast, so precision was required in this contest. As they approached the course, service spacecrafts that had been helping them prepare for battle and supplying the contestants with the fuel that they needed to get up to speed, collected the fuel tanks and peeled away, and the two combat vessels were alone. Spangled with electric lights so the cameras and billions of television viewers on both planets would be better able to see them, the spacecrafts were now headed almost directly toward each other and in perfect alignment with their places on the jousting course. Both crafts had turned off their main thrusters, and their attendants had cleaned away any detritus around them because the rules of the contest were that only laser beams were allowed to be used in this sporting event. The spacecraft were both traveling at 350 million miles per hour, that is to say, 93,000 miles per second, half of the speed of light! These speeds that the two vessels had attained were measured relative to the battleground, which they both were rapidly approaching, their timing perfect to the microsecond. Their trajectories, if plotted on a graph, would indicate a single straight line, so close were they calculated to pass by each other with only the motionless midcourse camera satellite between them. Now, both combat crafts were spotted by the camera satellites at either end of the 186,000-mile-long jousting course. These very high-speed cameras would provide slow-motion replay entertainment and provide the means for very careful analysis of what had taken place. The deadly laser projectors of these highly sophisticated and specialized fighting machines adjusted their aims fore and aft, the computers calculating where their opponents would pass. The two spacecrafts entered the starting gates simultaneously, their massive generators screaming, their laser cannons blazing hot ruby violence. Half a second later, the two crafts passed each other and the after-laser cannons opened fire. Half a second after that, the two crafts were out through the exit points and on their way home. Turning around for a rematch would be quite impossible. Beings on other planets too, throughout that galaxy would gasp, bubble or chitter in awe at the sight and eagerly await the slow-motion replays. What had happened? What did all of that laser light do? What did all of those cameras see?

Chapter Three

But What Would Really Happen?

According to the Special Theory of Relativity, the light went away from each contestant's spacecraft at the speed of light, relative to the fast-moving vessel. When a bit of the light struck the motionless satellite at the center of the course, it was traveling at the speed of light relative to it, too. (We'll say that somehow it didn't damage the camera. And who knows? Maybe it wouldn't have, as we will see.) When that laser light struck the other ship coming toward it going half the speed of light, again, it was going at the speed of light for them, too. After the two fighters passed each other, the lasers pointing opened fire, and light left them at the speed of light and hit the opponent at the speed of light. How did it slow down to the speed of zero going one way and accelerate to double the speed of light going the other way while still traveling at only one speed?

Einstein's explanation for this is that for both spacecrafts time is slowed down and space is contracted. Is that what happened, but then after they passed each other, suddenly time sped up and space expanded? Or was it the other way around? Can time and space adjust themselves so abruptly, in one way, relative to each other and differently relative to their surroundings? Or is it that time slows down, and space contracts both ways? (Then how would light know what to do?) Let's look at another explanation that satisfies the mystery of the speed of light without time and space having to do anything like that. Relativity can still work but in an easier way.

The stuff of light, photons, and waves of electromagnetic energy, comes out of candles, stars, lightbulbs and radio towers (not light but electromagnetic energy in longer wavelengths) at the speed of light. Is it that it comes out at no other speed? Why shouldn't there be more, lots more, of these quanta of energy/infra-subatomic particles, some traveling slower than, others traveling faster than the speed of light? The speed of light is not a barrier. Anything can travel faster than the speed of light. (Except, of course, that anything that ever did, hit something else long ago, and the difficulty of accelerating anything other than invisible subatomic particles and then having to interpret the results based on magnitudes of collision images.) Besides, relative to what? It's just that only the stuff of light that is traveling at a speed that is very close to 186,000 miles per second relative to the viewer can be detected in any way. Whether it's light or any other form of electromagnetic energy. At this "correct" speed of light, these infinitesimal particles/bits of energy don't gain mass, but merely incur some increased form of interactivity with local, relatively motionless matter and become palpable to the eye, instruments, or simply to molecules by striking them and increasing for a moment, the orbit of their electrons/Brownian motion—illuminating and warming them. The other slower than or faster than light-speed infra-subatomic particles/waves of radiated energy are the same as the photons/electromagnetic energy that you can detect, and if you were moving in a way that they were hitting you at 186,000 miles per second, then they would be visible photons for you. All the other particles from the same (or any other) source but moving at a speed different from 186,000 miles per second relative to you do some of the same things that they do. They may or may not pass through clear objects such as glass without electron orbit conversion. (If light is, as it is believed, slowed down/refracted because it goes through by expending itself to raise an electron to a higher orbit and promptly is recreated by the electron's fall back into a lower orbit and the light quantum going to the next electron then wrong speed light probably does not pass through glass because its unable to affect the electrons. But if refraction is a different and as yet unknown process, then maybe...) They bounce off shiny objects. They are probably absorbed by some objects. And, very likely, a large portion of them are not absorbed but bounce off of objects, having failed to interact with them but what? As reflected glare perhaps if someone looking at the object and moving at the right speed relative to it saw it. Maybe someday an experiment will tell the truth about that. But wrong-speed light will have no effect on any object—it is traveling at the wrong speed. How it would pass through glass without interacting with the electrons in the glass is a mystery to me; maybe wrong speed light would just bounce off or be absorbed. Or maybe it would pass through without interacting with electrons but just passing between them. This non-light might still be slowed down in passing through glass or any other clear but optically dense material. Maybe the electrons possess some force or action that slows but doesn't interact with wrong-speed light. Or maybe it just goes through at its same speed as though the atoms weren't there because their electrons have no effect on wrong-speed light just as it has no effect on the electrons. This wrong-speed light is so ephemeral and unaffected by the mysterious process that seemingly gives objects more mass as they approach the speed of light (but actually just makes them more interactive or more in step with some rhythm). This non-light might as well not be there. But it is and this stuff is everywhere throughout the universe. It's moving in all different directions and all different speeds, and I doubt that the particles that comprise this non-light ever collide or have any effect on one another. (Except perhaps between the two mirrors of a laser.)

One reason that it is easy to suppose that light would travel at one speed is because sound waves travel at the speed of sound. And like light, they travel at different speeds in different mediums. But sound waves are waves of the material of the medium itself where light waves are waves of light particles emitted by a combustive process. We see them traveling at one speed, and it's easy to assume that like sound, that is their only speed. We should consider, on the other hand, that just as mass (interactivity) of a particle increases as it approaches the speed of light, so should the "mass" of light itself. The increase in mass/tangibility of accelerated particles serves as a good indication of why we detect light only when it is traveling at the correct speed, that is, relative to us. More experiments directed at testing for the existence of "wrong-seed" light may turn up a new answers to a lot of questions.

Going back to red-shifting galaxies, is red shifting anything more than an electromagnetic doppler effect similar to change of pitch of propeller noise in the exciting part of a submarine movie? There is no universal medium in space to set the speed at which light travels the way water sets the speed of sound in the ocean. If those galaxies were somehow traveling toward us, even though the speed of their light is based on their motion, space and time would not have to change so that the light could slow down. We would see different light—the light that had left the galaxy traveling more slowly. It would be blue-shifted, not because it is coming at us squeezed by contracting space, but because it left going slower with the same frequency beating out shorter wavelengths. The light with the correct wavelengths now goes too fast for us and is invisible.

Let's get back to the dueling spaceships and take a look at what the viewers saw from a camera mounted on one of the spacecrafts that was pointed at the other. Again, each spacecraft is traveling around 93,000 miles per second (relative to the jousting course). Now, to please the audience, these spacecrafts have lots of running lights. In fact, let's say that they are lit up like casino signs. In a light bulb or other source of light, the projection of the stuff, particles/waves of light that can travel at any speed, can be compared to candy knocked out of a piñata when the batter hits it. Some just falls close-by, and some flies far away. Likewise, on our conveniently, brightly lit spacecraft, some small amount of the light comes out of those LED bulbs (I suppose that's what they would use) at a snail's pace, and some small amount comes out at a speed far faster than the given speed of light. And then, in addition to this light, since there is already light, mostly from stars, in space moving at many different speeds and in all directions, there is presumably always some light from some source that is moving at the right speed and direction to strike the spacecraft. It's just that this light, at the time that it hits the spacecraft, is not traveling at the correct speed of light so as to illuminate the ship and come glowing off of it having inter-acted with the electrons on the metal hull and be traveling at the speed of light relative to the moving spacecraft so that another craft traveling along-side would see the spacecraft even more clearly. This other-source, different-speed light can also be re-flected off of the shiny metal hull and be traveling at its previous speed plus or minus the speed of the spacecraft. It would not have interacted with the spaceship, not have elevated any electrons to a higher orbit and then been recreated by that electron falling into a lower orbit. This light was wrong-speed for a camera mounted on the spacecraft to see and is still wrong-speed for any party traveling with the spacecraft to see. Now maybe only some of this light, only the reflected glare is now traveling at the right speed to be detected by the camera on the other spacecraft com-ing toward it or away from it.

But what about color-shifting? If the speed of one object is so wildly divergent from its viewer, can it shift right out of the visible spectrum? Maybe yes, but let's just assume for convenience's sake that there is always electromagnetic radiation from x-ray frequency to radio wave frequency coming from or bouncing off the object viewed to be now shifted to a visible frequency as well as be traveling at the right speed and direction to be seen by a viewer moving very differently. So, under laser jousting circumstances, color shifting may be extreme, but let's assume that it can be coped with. Maybe the advancement of photo-technology in this galaxy is so great that cameras detect all sorts of electromagnetic radiation and automatically convert it or correct their portrayal of it to an acceptable rendition of visible light. Another question comes to mind. If wrong-speed light cannot interact with matter—illuminate, heat it, pass through glass, be seen or detected—then how is it created? I have to think that most light created in the inferno of a star, candle, or light bulb filament, is right-speed but in that chaotic frenzy of electron (or nuclear) movement, some wrong-speed light is created too.

During the time that the two spacecrafts are approaching each other, since both spacecrafts are going at half the speed of light, only the light that is leaving either spacecraft very slowly will hit the other at the right speed to be seen, be otherwise detected, or to have any effect. The combatants (the computers) will be firing blind. They will be using very sophisticated game-theory-generated algorithms to guess where the other craft is and similar algorithms to not be where the other computer is likely to guess where they are. And, of course, to not cross some theoretical line and be in a really bad place. (Crash!) They will only see each other at the last moment, just for an instant as a long, stretched out persistence of vision, an instantaneous streak of all the light that left the oncoming spacecraft so slowly that it piled

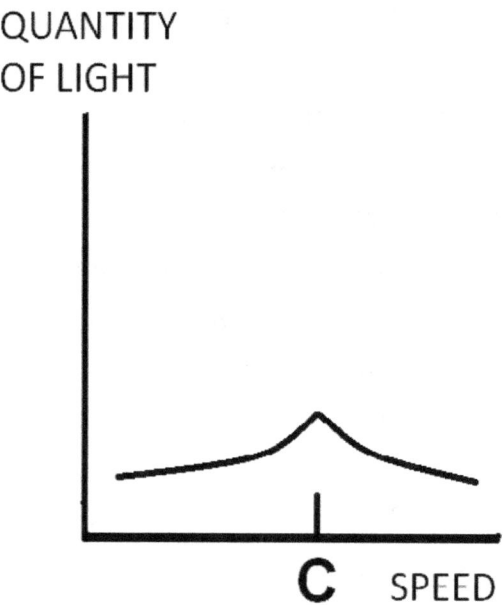

Bell Curve Showing the Amount of Light Going Faster or Slower Than the Given Speed of Light

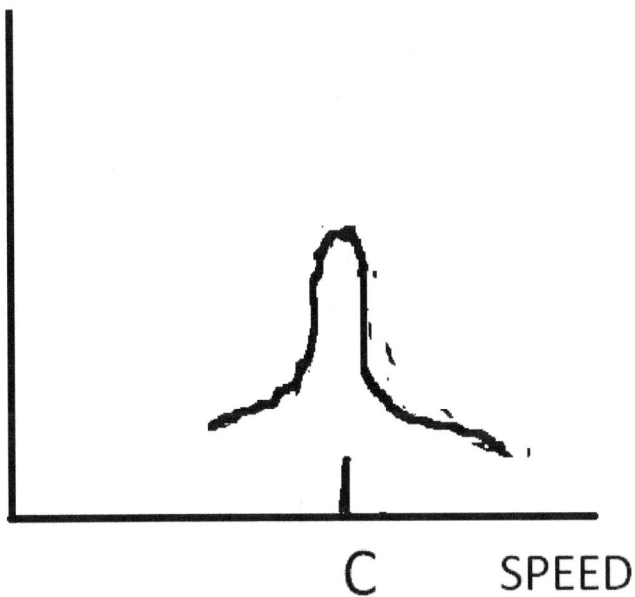

Bell Curve Showing the Amount of Laser Light Going Faster
or Slower Than the Given Speed of Light

up in front of it. And both competitors will see this flash, this long, stretched out blur only at the instant that they pass each other. It will look to both of them as though the other was traveling far faster than the speed of light. They will not see what actually happened; that the other ship came through the gate and then took a whole second to reach them. Only light that was coming out of their laser gun at near zero speed would have any effect on the other spaceship. Does any of the stuff that comes out of a laser travel that slowly? Does a laser's highly concentrated beam of light that has been bouncing back and forth between two high-efficiency mirrors tend to beat faster-moving and slower-moving photons into moving not only in frequency lockstep but also to an average speed that just happens to be very close to 186,000 miles per second and so increases the volume of palpable right-speed light? If you plotted on a graph, the amount of all the particles coming out of an ordinary light source (not a laser), from slower moving on the left to faster moving on the right, that is, the x-axis showing the speed and the y-axis showing the quantity of light, the figure displayed would be a bell curve centered on 186,000 mps, the given speed of light. Would the bell curve of light from a laser be taller and narrower than the bell curve of ordinary light? Someone will someday perform an experiment to determine if this is so.

Chapter Four

More on Optical Illusion Relativity and Different Speed Light

What do you see from the rear-pointing camera of one spacecraft when you watch the other one zoom away after the two ships have passed? Now you see only the light from that spacecraft that came (in your direction) going 372,000 miles per second, double the proper speed of light when it left the other spaceship. It needs to go 93,000 miles per second, the other ship's speed plus 93,000 mps, your speed plus 186,000 miles per second to be going the speed of light for the other spacecraft's camera. One second after passing each other, when both spaceships are passing through their exit gates opposite the ones they came in through, 186,000 miles apart, the 372,000 mps light from one spacecraft (that is traveling 279,000 mps relative to the jousting course) will take a second to get to the other spacecraft. But in that one second, the observing spacecraft will have travelled 93,000 additional miles while the 279,000 mile per second light (relative to the course) travelled the length of the course plus the extra 93,000 miles to catch up with the observer. So, since the watching spaceship saw that it took the other spaceship two seconds to reach the opposite gate after they passed, it appears to them that the other ship was going only half of its actual speed; that is, 46,500 mps. What a sudden (apparent) slow-down! And it looks the same way to both of them in regard to the other because each craft is moving away from the other just as much as the other is moving away from him. They are both still going just as fast as before they crossed paths. At their high speeds and great distances, there is naturally a gap between where they are and where their image is that the other ship sees at that moment. But now that it is fast-moving light from each that the other sees, those gaps are smaller. The contestants see each other continuously, so the enormous apparent change in speed of the other looks sudden but smooth and seamless. How can the change be so great? Let's break it down into two parts and see. The apparent change from infinite speed to normal speed is much greater than the apparent change from normal speed to half speed. This is because the percentage change of near-zero speed light to normal speed light is greater than the change from normal speed light to double speed light that is seen coming from the other spaceship. When looking behind at the other ship, each observer is also looking at a jousting course that they are moving away from at high speed. The light coming from that course is coming to the receding observer at one-and-one-half times its normal speed. No matter—the length of the course is already established. If that fast-moving observer imagines that he is standing still, then he will perceive the course to be moving. He will see it moving away more slowly than it really is, and he will see a distortion of its size or shape. The length and shape of the jousting course is known and whatever deforming of the course that the observer sees, he mentally adjusts to what he knows it to be. When there is no reference like a jousting course, it gets trickier. We will see this in one of the calculations in Chapter Six.

What would the audience see from the camera in the motionless satellite at the middle of the course watching the two ships speeding away from it? The camera would not see the same light as the cameras mounted on the contestants' ships saw when the ships were going away from each other. That light left each ship going 376,000 mps. The camera in the middle would see the light that was leaving either spacecraft at 279,000 miles per second, so to the camera's audience, they would appear to be traveling at two-thirds of their actual speed, not one half. This light, going 186,000 miles per second relative to the course, from one of the spaceships passing through its exit gate, travels 93,000 miles to the middle camera and takes one half of a second to get there. By the time the camera sees that the spacecraft has gone the 93,000 miles from the midpoint to the exit gate, it has gone an extra 46,500 miles, one and a half times the distance that the middle camera saw it go. It appears to be going slower, 2/3 of its actual speed. (That is, one third of the speed of light.) What about from the time a spacecraft passes through its entry gate to when it passes the motionless camera in the middle? The light leaving it at 93,000 mps hits the camera at 186,000 mps. The light from the spacecraft at the gate takes one half second to get to the camera; the spaceship takes one second, so it appears to have gone the distance in half a second. The midcourse camera saw all the travel from gate to beside it in the latter half of the second that it took the spaceship to get there. It appears to be traveling at the speed of light, twice its actual speed.

Chapter Five

Not Only Does Time Not Slow Down,
but Space Doesn't Contract Either

Let's take a different spacecraft, one that is 93,000 miles long (yeah, yeah, I know, but this is a hypothetical spacecraft) and going away from us at 93,000 mps. Let's say that we are watching it from a motionless camera at its entry gate at the far end of the course from where the nose of this needle-shaped craft is going out through its exit gate. (And as its tail is passing the midway satellite.) Again, light that our motionless camera could see would be leaving this spacecraft at 279,000 miles per second but would be traveling at 186,000 mps relative to the course for the audience to see it. The light from the back end of the spacecraft takes one half of a second to go from the midway point to our camera at the entry gate. But to go from the midway point to our camera at the entry gate along with the image of the back end of the spacecraft, the image of the front end of the spacecraft had to leave the nose one-third of a second earlier. At the end of that one-third of a second, the light from the front end of the ship will have travelled 62,000 miles to the midway point just as the back end of the ship arrives at the midway point and light from the back end and the light from the front end, both together (with the whole image of the spacecraft between them), leave from the midway point heading for the camera behind them at the entry gate. But one-third of a second before that light left the midway point, the nose had not reached the exit gate. It was 32,000 miles short of it when light left it and went 64,000 miles in one-third of a second back to the midway point where the tail had just arrived. So, the space-ship appears to the ever more distant but motionless camera to be 64,000 miles long, two-thirds of its actual length. And likewise, it took the tail one second to get from the camera to the midway point and one half-second for its image to get back from the mid-way point to the camera—it takes one and a half times as long for us to see it get where it's going—it appears to be going two-thirds of its actual speed.

Not only does it look like the spacecraft is going at two-thirds of its actual speed but it appears to be only two thirds of its actual length. (This optical illusion would be the same for shorter space-crafts. I just use a long spacecraft for easier illustration.)

What if this 93,000-mile-long spacecraft were coming toward our motionless camera at 93,000 miles per second? Our camera is at one end of the course, and the nose of this slender spaceship has reached the midway satellite, and its tail is at the other gate. Light from this spacecraft will have to leave it going half the speed of light to be going 186,000 mps to our motionless camera. One-second before the tail of the spacecraft got to where it is now, at its entry gate, that tail was 93,000 miles back of that gate and light from it started on a journey toward our camera and went the 186,000 miles to the midway point in one second to get there just when the nose arrives. Now both images leave the midway point, and one half a second later, the camera sees a 186,000-mile-long spaceship approaching. Half of a second after that, the nose is right alongside the camera; it appeared to have travelled 93,000 miles in half a second. It appears to be twice as long as it really is and to be going twice as fast as it is actually going. Time and space don't have to change for us to see the strange effects of relativity; the optical illusion of relativity is made possible by light going at different speeds, based on the speed at which and the direction that the object we are viewing is moving, relative to our motion.

Chapter Six

Quantifying Relativity

Let's look at some equations that will help us calculate rates of apparent speed to real speed and apparent spaceship length to real spaceship length. Our first term for velocity, Va represents apparent speed; Vr is real speed. If both spacecrafts are moving toward or away from the other, then we will add another small letter to the "V" and we will have Vac or Vrc. (If an observer is chasing and catching up to the observed, they are coming toward, but we subtract the lesser speed. If the one being chased is escaping, we say going away but subtract chaser's speed.) La and Lr have the same meanings for the observed spaceship's apparent and real lengths. We will again, combine the velocities for the speed factor in the equation. There is only the observed spaceship's length involved. C is the given speed of light, but since we are going to be giving our answers in fractions and multiples of the speed of light, let's just usually think of C as "1". Now light travels at all speeds and only light that hits the observer's eye or camera at 186,000 miles per second is seen. (The speed of light has been given to be

186,172 miles per second, but to keep the arithmetic easy, I will round it off in this book.) We will have to use both the normal speed of light, and the speed of light plus or minus the observer's speed going away from or coming toward the observed. What about the performer's motion? It adds no momentum to light. At any instant the point where light left the observed is motionless. Although for velocity, both observer's and observed's speeds are combined, we add or subtract only the observer's speed away from or toward that motionless point to determine what fraction or multiple of C we will use for the speed of light that is observed. Let's call this speed of the light that is observed "Co".

Coming toward us:

$$Vac = Vrc \times Co / (C - Vrc)$$
$$La = Lr \times Co / (C - Vrc)$$

Going away from us:

$$Vac = Vrc \times Co / (C + Vrc)$$
$$La = Lr \times Co / (C + Vrc)$$

In the laser jousting contest, our two contestants were each going only half the speed of light but combined, their speed toward each other was the speed of light. But in other cases, observer and observed may be going away from each other or both going the same way. What then?

The following equations show how when one spaceship is chasing another, the leader going half of the speed of light and the follower going one quarter the speed of light, appearances can be complicated. Let's take a case where the chaser sees how fast the quarry appears to be going but doesn't actually know. To switch Va and Vr and solve for the real speed, all we have to do is switch the two speeds of light putting Co in the denominator and C in the numerator, and we have to switch the arithmetic sign in the parentheses, plus to minus or vice versa. So, let's say that a large, heavily armed spaceship is chasing a smaller, less heavily armed one through a dusty part of the galaxy. If the large spacecraft goes faster than a quarter the speed of light, the dust overheats its protective carapace. The small, slim, and well-protected spacecraft can manage half the speed of light without damage.

$$Vac = Vrc \times Co / (C + Vrc)$$
$$Vac = 1/4 \times 3/4 / (1 + 1/4) = 3/20$$

The above equation shows that the apparent difference in their speeds is three-twentieths of the speed of light. That is to say, it shows their combined speeds relative to each other. The speed of the slower one is subtracted from the faster one so that we can see how fast one appears to be going away from the other. Let's say that the captain of the chasing ship wants to know what that speed is:

Captain: Chief, how fast are they going?

Chief Systems Engineer: Captain, our optical range finder indicates that they are going three twentieths the speed of light faster than us.

But now, the executive officer intervenes to clarify what that reading means. He does a quick mental calculation:

Vrc = Vac x C / (Co - Vac)
Vrc = 3/20 x 1 / (3/4 - 3/20)
Vrc = 3/20 x 5/3 = 1/4

> Executive Officer: Actually Sir, they're getting away from us at a full quarter of the speed of light. Our going one quarter of the speed of light is playing tricks on our range finder.

Going back to the two laser jousting contestants, at the moment they met each other in passing and saw what looked like each other's infinite speed, they knew how fast the other was going because:

Vrc = Vac x C / (Co + Vac)
(It's a plus sign because both, they're coming toward each other and we are deriving the real velocity not the apparent velocity. A negative multiplied by a negative is a positive.)

Vrc = Inf x 1 / (1/2 + Inf)

Eliminate the 1/2 in parentheses; it's insignificant compared to infinity. Divide the Inf in the numerator by the infinity in the denominator and:

$Vrc = 1/1 = 1$

If each contestant's speed is half of the speed of light and they calculate that their actual combined speed is the speed of light, then
they each know that the other contestant's speed is half of the
speed of light.

Going away from each other after they pass, they each see
each other's speed as:

$Vac = Vrc \times Co / (C + Vrc)$
$Vac = 1 \times 3/2 / (1 + 1) = 3/4$

Without a jousting course beside them, each spaceship might perceive itself as motionless and simply see the other going away at
three-quarters of the speed of light. Now remember that when
the one had gone out his exit gate and another 93,000 miles, he
looked back and saw that the other had only reached his exit gate.
They saw that together they had separated by only three 93,000
laps, not the full four that were between them. The jousting
course is not moving, so its length is not distorted. With or without the course, their apparent combined speed is three-quarters
of its real speed.

So, what is the apparent length of the jousting course to that
spaceship speeding away from it at half the speed of light?

$La = Lr \times Co /(C + Vrc)$
$La = 186,000 \times 3/2 / (1 + 1/2)$
$La = 186,000 \times 3/2 / 3/2 = 186,000$ miles

There is no change in the shape of the course for a fast moving observer.

Let's suppose that a spaceship is going away from our motionless camera at infinite speed. What would our camera see? (No need to look for a combined speed or to subtract our speed from the speed of the light coming from the spaceship.):

$$Va = Vr \times C / (C + Vr)$$
$$Va = Inf \times 1 / (1 + Inf) = 1$$
(We can say this because 1 + Infinity = Infinity.)

It seems to be departing at only the speed of light.

So, it is (only theoretically until something does it) possible for objects to travel at the speed of light or faster. It's just that when they approach an observer at the speed of light, they appear to be doing something that is impossible, that is, going infinitely fast. Going away on the other hand, they will have to go infinitely fast to appear to be going the speed of light.

Taking a look at our illustration on a nearby page, the real vs. apparent velocity graph, it is to be understood that real velocity would be a straight line coming out of the origin at a 45-degree angle. The illustrator left it out to avoid clutter. The line representing apparent speed for coming toward the observer would leave the origin at a 45-degree angle since at slow speeds real and apparent speeds are the same. The line would curve and rise steeper and steeper as it approaches the (not shown) vertical line that would be starting at "C", on the x-axis. This absent line be

REAL VELOCITY VERSUS APPARENT VELOCITY TO AN OBSERVER

COMING TOWARD

$$Va = Vr \times C_o / (C - Vr)$$

GOING AWAY

$$Va = Vr \times C_o / (C + Vr)$$

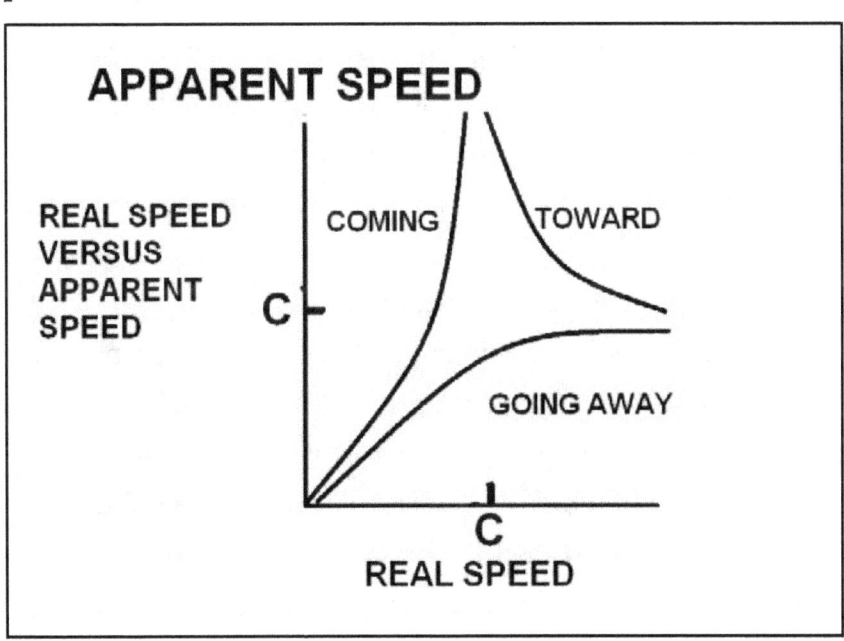

APPARENT SPEED

REAL SPEED VERSUS APPARENT SPEED

COMING   TOWARD

C

GOING AWAY

C

REAL SPEED

tween the two asymptotic lines running up to infinity is the given speed of light. After exceeding the speed of light, we see the other side of this line representing "object and viewer's coming toward each other" line of apparent velocity again. Now, it is coming down from infinity and asymptotically moves away from the "C" line going up from the x-axis. Then, taking the shape of a hyperbolic curve, it will approach the other C-line on this graph; that is to say, the speed of light's horizontal line coming out from "C" on the y-axis. So, the faster it goes after exceeding the speed of light, the slower it appears to be going but never appearing to go as slow as the speed of light again. We will see more about this strange turn of apparent speed in Chapter Eight.

The other line on this graph, lower down, is the "Apparent Speed When Object and Viewer are Moving Away From Each Other" line. It comes out of the origin and curves toward but never quite reaches flatness. It shows, for example, how at C on the real speed x-axis, the object appears to be going halfway up to the horizontal C line coming out of the apparent speed y-axis.

As this away from line approaches real speed infinity, it moves up closer to apparent speed of light C.

For the real vs. apparent length of the spaceship graph, coming toward, the line representing actual length would be a horizontal line representing the true length of the spaceship. Let's say starting at the point on the y-axis representing 93,000 miles long (for one of our longer than 11 Earths lined up side by side superspaceships.) The line of apparent length of such a spaceship coming toward the observer would start out following the horizontal line of real length and then would curve upward. Then it would

REAL LENGTH VERSUS APPARENT LENGTH

AN OBSERVER:

COMING TOWARD
THE OBSERVER

$$La = Lr \times C_0 / (C - Vr)$$

GOING AWAY
FROM THE OBSERVER

$$La = Lr \times C_0 / (C + Vr)$$

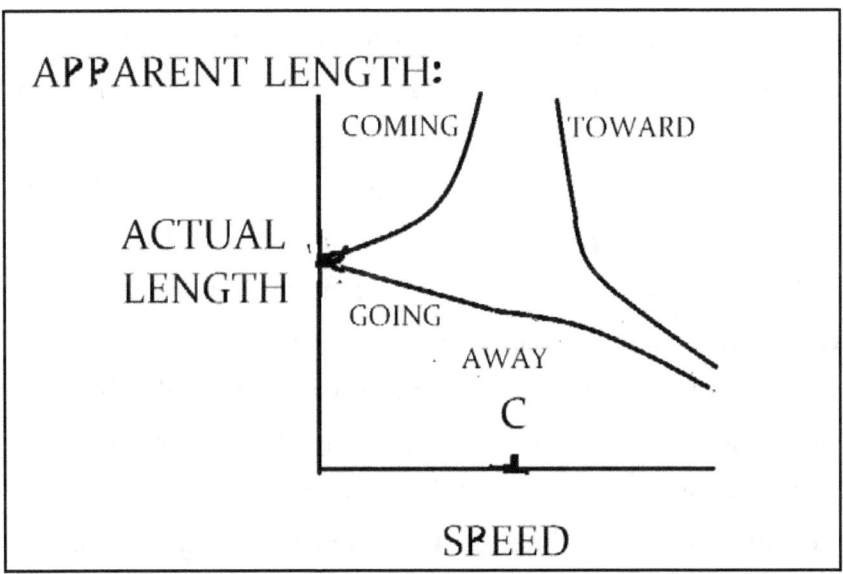

A Graph Showing Real and Apparent Lengths of Objects Coming and Going

do the same as the velocity line until going faster than the speed of light. After the line representing apparent length gets past the "C" value on the x-axis, that is to say, the speed of light, it would come down from infinity the same way that the apparent speed coming toward line did, but it would go down past the horizontal line representing the actual length and approach zero as the spaceship's real speed approaches infinity.

The lower line on this graph, representing apparent length of a spaceship going away from the observer, starts out following the horizontal line of actual length and curves down to approach but never reach the x-axis. The line shows that as the spaceship's speed approaches infinity, its apparent length approaches zero.

What would a camera see if it was one million miles away from the jousting course and perpendicular to the course at its center point. For practical purposes, all light from along the course would reach this observing camera's very powerful telescopic lens at the same time after a 5.4 second delay. That camera would see everything the way it really was happening. No relativity effect; it's seeing all the action on the jousting course at a right angle and the spacecraft is neither moving away from the camera nor toward it. Let's think of this arrangement as the triangle of a trigonometric diagram drawn on a graph with an x- and a y-axis. The camera observing from one million miles away is at the origin where the x- and y-axes intersect. The line of view from this camera to the satellite at the center of the course and to, let's say "M" for "million mile mark" on the x-axis is the run going along the x-axis. The path of the spacecraft goes from the jousting course's midpoint to the little image of a space craft and

RELATIVITY APPROACHES
FULL
EFFECT AS ANGLE OF VIEW
APPROACHES 90 DEGREES
RELATIVITY EFFECT EQUALS
THE SINE OF VIEWING ANGLE
(THAT IS THE RISE OVER
THE HYPOTENUSE)

RELATIVITY EFFECT APPROACHES ONE FOR ONE AS ANGLE APPROACHES 90 DEGREES.

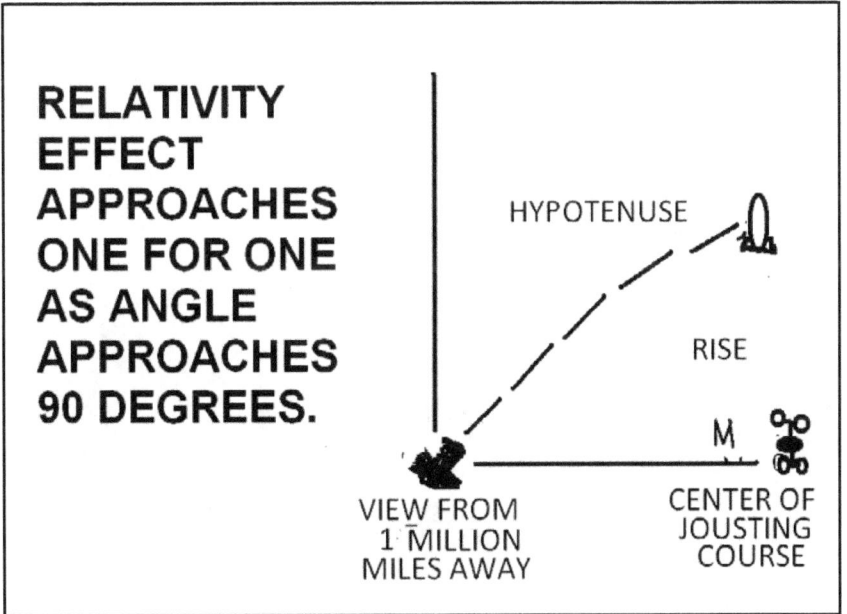

HYPOTENUSE

RISE

M

VIEW FROM
1 MILLION
MILES AWAY

CENTER OF
JOUSTING
COURSE

it shows the rise of this right triangle. Finally, the line of sight from that far-away camera at the origin of the graph to where the spacecraft has gotten to is the hypotenuse. As the camera's angle of view swings from the spacecraft's start at the center of the jousting course (and the million mark on the x-axis) to where it is getting to, that is to say, as the rise, increases, the relativity effect would increase from zero to the sine of the angle that is formed between the run and the hypotenuse. That is, the ratio of the rise over the hypotenuse. After this spacecraft has travelled many millions of miles, the relativity effect will have gone from zero to nearly 100 percent.

Chapter Seven

Relative to This or That?

Now about the Michelson-Morley experiment. In the late 19th century, these two men demonstrated that the distance light travels remains the same, unaffected by which direction it is going. This seemingly showed that no matter what direction the earth was moving through space or how fast, light still traveled the same distance on their light-distance-travelled-measuring table. Using a semi-transparent mirror set at a 45-degree angle, they split a beam of light 90-degrees apart, bounced the two light beams off of two mirrors, and back to two vertical slits to a diffraction image that would catch the slightest variation in distance the light traveled. No matter which way they turned the very stable apparatus, there was no change in the distance that the two split beams of light traveled. The Earth is certainly in motion so this was a troubling outcome. Whatever way the Earth was going, wouldn't the difference in the way the beams had to go relative to the Earth's motion make some difference in how far the beam actually traveled? If light left its source on a moving Earth, and the target was riding on this moving Earth toward or away from the source, wouldn't the distance that it had to travel be changed at least the distance of a light wave? The solution was put forward by Albert Einstein that since light travels at only its one speed, space must contract and time must slow down to enable light to ignore the motion of its source and target.

The problem is that in 1908, George Sagnac did a similar experiment in which the distances light travelled changes. I will talk about the Sagnac device further along in this book, but the fundamental differences in the two experiments was that the Michelson-Morley experiment was measuring the distance light moved between moving mirrors and depended on Earth's motion through space to do the moving. Sagnac was not measuring how far mirrors moved away from or toward each other—he was only using them to conduct light between a moving source and target. He did not depend on Earth's motion through space either. He moved the source and target himself. This showed that more experiments are needed to explain the different results. I hope someone will use the wheel of mirrors in Chapter One and another experiment using a laser on a mountain experiment that we will get into further along in the book, to find out more about what moving mirrors do to light and whether Earth's motion can be measured here on Earth using light. Let's use spaceships as our model to replicate and take a different look at the Michelson-Morley experiment.

Three spaceships are moving through space in precisely the same direction. Two are traveling abreast of each other, 186,000

No matter which way the beams are turned the pattern of light and dark bars on the display card remains steadfast.

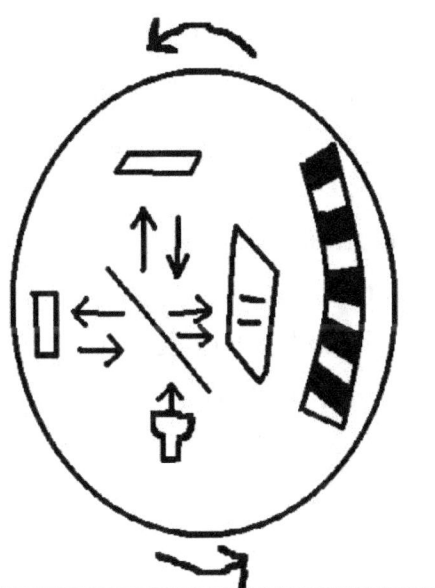

The Michelson Morley right angle path interferometer shows that no matter which way the apparatus is oriented, the general motion of the earth's rotation or travel in space does not shorten the path that either beam of light takes to the difraction grating.

The light from the source hits a half-silvered mirror. Some passes through and some bounces off at a 90 degree angle. Both beams are reflected back by two mirrors and some of the light that passed through is reflected to the two slits of the diffraction grating and vice versa for the other beam.

miles apart, and going 93,000 miles per second. The third is going the same speed and direction as the others, but it is 186,000 miles ahead of the one on the right. Let's call the leader A, the spaceship on the right B and the one on the left C. B has a bright light that shines at A and C and can be flashed on and off. This light is not focused on A or C; it shines broadly around A and C and between them. Michelson and Morley must have used a pretty broad beam of light to hit both slits in the diffraction card and the mirrors that they used were not tiny and the distance, diagonal beam splitter to mirrors to target was only a few feet so the direction that the light went was not an issue in their experiment either. Only the distance that it traveled. (On the other hand, if I were in Spaceship B, I would want to shine a laser, not a harmful one, at C to see if B's momentum affected where the light went. Would the beam go just where it was aimed relative to background stars or to where Spaceship C had gotten to when the ray of laser light arrived one second later courtesy of B's forward momentum?) A and C have big mirrors pointed at B's light. I'm afraid that it's not good enough to just say that the mirror is retroreflective, so we'll say that it is being controlled to make sure that it reflects the light back to B. And we will see that generally that for Spaceship C, "aimed to reflect back to B" means that it is left simply parallel to their direction of flight. Mounted on Spaceship B next to the light but shielded from most of that very bright light is a light detector with a timer. The timer, like a stop watch, measures the time elapsed between the flash of the light and the flash of its reflection. The detector's aperture is pointed at A and C, not at the light beside it but it can detect both flashes.

They all fly so steadily that the distances between the mirrors and the light never change one bit. Spaceship B flashes its light. Let's take C first and we will start by disregarding the spaceships' mutual motion and consider it the same as though they were standing still in space. Light leaves B at 186,000 miles per second, bounces off of C's mirror (which is neither moving toward or away from the source of light), returns at 186,000 mps, and two seconds later, is detected by the timed sensor on B's nose. But we must consider the 93,000 miles per second speeds of B and C. The motion of a body is not added to the light emitted from it, and we are going to consider the light coming out of the source that is aimed to hit the mirror where it will be when the light gets to it. This light is aimed up toward the way that the mirror on C is going at an angle of a little more than 30 degrees. (The tangent of this angle equals one half.) Light simply goes where it is aimed and not with any added impetus toward where its source is going. (Unless someday an experiment proves me wrong.) The x-axis speed of the light is still 186,000 miles per second. Its y-axis speed is 93,000 miles per second. It will bounce off the mirror and arrive at the sensor two seconds later, the x value of its motion will be 186,000 miles per second, and the y value will be zero relative to the sensor. (Spaceship B and the light both have a y value speed of 93,000 mps that cancel each other out.) The actual speed at which the light traveled will have been the square root of the sum of the squares of 186,000 mps and 93,000 mps. That is to say, about 208,000 mps. But only because all of the vast amount of light, seen and generally unseen, that came out of the search light was traveling at lots of speeds and that particular speed, 208,000 miles per second, enabled the sensor to detect it going 186,000 mps.

For Spaceship A, if we just think of A and B as motionless in space as well as relative to each other, then the light takes a simple 186,000 mps, two second trip as it did for B to C and back when we didn't consider their motion. If now, we consider the A and B spaceships' motions, then it is light that left B at one and a half times the speed of light (relative to the point in space where the light left B's search light) that will be operative. Not because we must add the speed of Spaceship B to the usual speed of the light but because only light going at that speed will get back to the timer/sensor in the right amount of time and traveling at the right speed to be detected. It will reach A's mirror in one second although it had to travel 279,000 miles. Relative to the motionless space around them (or more aptly, from the motionless point from which the flash of light left the searchlight), the mirror is receding from the light at 93,000 mps. The light loses 93,000 mps in hitting the retrograde mirror and then loses the same amount of speed AGAIN in leaving the backward moving mirror! (I hope somebody does my Chapter One wheel of mirrors experiment or one like it real soon to check if I'm right about this.) It will return to the sensor on B's nose at 93,000 mps and will go 93,000 miles to reach the oncoming sensor. The sensor records that the light made the round trip in two seconds. (It could be that a mirror simply remixes speeds of light for this to happen. Or perhaps this experiment is just a little more demanding than Michelson-Morley's regarding speeds and light would never get back to the sensor.) So whether we consider the light to be going perpendicular to the spaceships' vectors (B to C) or parallel (B to A) and whether the spaceships are motionless or moving, the light always makes the trip in exactly two seconds—a result similar to that of the Michelson-Morley experiment.

But still, how do we know what their motion is relative to the space around them. Not by bringing "aether wind" (an imaginary medium everywhere in the universe in which light is supposed to create light waves) into this issue, but which way is "motionless" space going relative to them? We say, "The motionless point where light left its source," but which way did that point have to be going to be motionless? Let's suppose that the astronauts in these three spaceships were traveling between galaxies and they were in a part of space far from any stars or galaxies. They have no reference points, but facing a long journey, they climb into their hibernation chambers for a long snooze. While they are asleep, the computers (each one referred to as Systems Control) on all three ships enter into a conspiracy led by the computer on ship B. The astronauts have nicknamed him SYCO. The three computers play a trick on the astronauts and radically change the arrangements of this journey. One of the astronauts in Spaceship B named Dudley wakes up before the others. He looks out a porthole and sees that Spaceship C is not 186,000 miles to their left and 93,000 miles behind (which is where it should appear to be since light takes one second to cover that distance and the ships cover 93,000 miles in a second). Spaceship C simply appears to be 186,000 miles to their left and beside them. Is it that C's speed has crept up and it is now 93,000 miles ahead of where it should be? Is C's image going directly to the right now and hitting B one second later when B is alongside the spot where C's image left from?

Dudley: What's going on, SYCO?

SYCO: That's not for you to be concerned about, Dudley.

After a fruitless conversation with SYCO, Dudley looks at the instrument panel, but it is dark. SYCO has turned off all of the instruments. Dudley soon realizes that SYCO and his cohorts are completely in control. Still, he wants to know what progress they are making on their trip. He puts on a spacesuit, grabs a big laser (not a blaster, just an illuminator) and climbs into the spaceship door's pressure lock chamber. He closes and seals the inside door, lets the air out of the chamber, and opens the outer door. He looks across and sees Spaceship C. Assuming that C is 93,000 miles ahead of where he sees it, he points the laser to a point 186,000 miles ahead of where he sees it. He is aiming it where he presumes that it will be when his laser light gets to it, switches the laser on, and rapidly glancing back and forth, he makes sure that he's pointing the laser to where C will be in one second while he watches the visible spaceship to see if a bright red dot flickers across it. His timing and his aim would be correct if the three ships were going 93,000 miles per second, but C across the way is not showing a laser dot dancing on its hull. (Er, he has a pair of very powerful binoculars built into the helmet of his spacesuit and these are big spaceships.) Dudley moves the beam further forward and sees no laser dot on C across the way. He then slowly, aiming up and down, works the beam back until, finally, he is pointing it directly at C's image abreast of B. There's the red dot on C! Dudley realizes, to his horror, that SYCO has stopped the little convoy dead in its tracks and far from any help. They are motionless.

So maybe if we could aim a laser from one high mountain top to an array of photosensors on another about 20 miles to the north (or south) and choose these two mountains from ones near the equator then the Earth's rotation would give us an opportunity to actually see a variance of light's motion caused by Earth's motion through space. If the rotation speed 1,000 miles per hour one way and 12 hours later, the same speed going the other way matters, then the variances in the red spot's position on the array would expose the motion of the Earth. The Earth's rotation would be perpendicular to our north-south arrayed mountaintops so our interest on the movement of the red spot would be mainly left to right. (I wonder how fast our galaxy is moving and which direction, but it would take too long to find any variation to reveal it. We want to just see if our red dot moves back and forth about three inches in a 24-hour period.) We would build shelters for the laser and the target but whether from wind or seismic motion, there would probably be some jiggle of the spot on the target. A computer connected to the sensors and programmed to average out the places the spot went to a single spot would likely solve that problem. Every 12 hours, we would see a variance caused by the change of the array's direction from the rotation of the Earth. (A similar variance would occur every six months for the Earth's motion around the Sun but that would not interfere with our experiment.) Since we know the timing of peaks of the variance back and forth and the speed involved, then despite the difficulties and large proximities involved in this experiment, we could confirm that light goes where it is aimed regardless of the motion of its source. So not in trying to find the Earth's motion by timing the speed of light moving relative to the Earth, but by measuring Earth's lateral motion during the light's passage, we would find out which way Earth is traveling through space. Or at least we could confirm that which we already know, the speed of its rotation, say, at the equator.

This experiment is important for another reason, too. I'm not really sure that light doesn't gain momentum from the motion of its source. I assume that it doesn't—it is a wave of energy, not a projection of matter. But in doing this experiment, would that spot of laser light move three inches from left to right every 12 hours and back again the next 12 hours? If it did so and indicated that a source's motion has no effect on how light moves, then does the motion of a mirror have any effect? Would my wheel of mirrors in Chapter One accomplish nothing? In my example about two spaceships going half the speed of light, one 186,000 miles ahead of the other and the one behind shines a light on a mirror on the other, would the light detector on the following one have to wait a long, long time for half-speed light to get to the mirror moving away at half the speed of light so that it could bounce back at half the speed of light so that he could see it at the speed of light? This being because the mirror's motion has no effect on the speed of light bouncing off of it. The truth is that no one can build hypotheses without extensive experimentation. One experiment is not enough. (And the experiments seem to just get more and more difficult and expensive.)

Let's take a closer look at that laser jousting spacecraft that was coming at his opponent at the full speed of light because their combined speeds came to that. How each contestant would be invisible to the other until they reached each other's camera and then would seem to burst out of nowhere. Their images would appear to be endless for only an instant. But which one was actually moving faster? It doesn't matter much here; the equation and the results are the same, but how do we establish a reference point? We could use some nearby star that is so massive that how could it be doing anything other than sitting still. (Okay, ponderously whirling around a galaxy at a moderate speed.) If you approach it at a little less than the speed of light, then you will see its slow-moving light well in advance enough to avoid flying into it. If two spacecrafts are flying toward each other at combined near-light speed and there is no handy reference star nearby, then no matter which is moving faster, one's slow-moving light gets to the other just as fast as the other's. But if a spacecraft is approaching a star at full light speed then theoretically, he would not see any zero-speed light and fly right into it unaware of its presence. (On the other hand, there's all that more substantial solar wind that the spacecraft will be hitting to warn him.)

Now everything in space is moving—even stars—so as a spacecraft going a little less than the speed of light approaches a star and sees newer and newer slow-moving light, he might notice that the star is going somewhere a lot faster than the star really is moving. But to be able to do this, he would have to be very aware of his own true motion. Because the star is so big, bright, and its motion is steady and the same as any nearby stars, he will perceive it as a reference point and see himself approaching it at near the speed of light. The star's motion around the galaxy, even speeded up by his high-speed approach, he will likely impute to his own motion. "The star isn't moving away to the side, I'm just going off to the other side of it not quite as fast as I could be going." This illusion that we're calling relativity only influences how each party sees the other party's actual motion and then it could seem to be their motion or his own to him. He may use a nearby star as a reference point or think of himself as stationary and consider the other to be doing all of the moving. And why shouldn't he be right whatever he thinks? It doesn't matter though, because it is the combined speeds that give the visual distortion results and who is doing the most moving just determines who will do more correcting. To interpret what they see, to find the distortion, two moving spaceships watching each other, must calculate the other's speed relative to themselves, not a reference point.

From any given start time to finish time, it is the other party's motion that is exaggerated if we consider ourselves to be watching from a motionless reference point. But if we think of ourselves as seeing the spacecraft's motion from an observing and moving spacecraft then our motion has to be included in calculating what distortion we will see.

What if a spacecraft is 93,000 miles away from its gate to the jousting course going at half the speed of light and at the other end is a 93,000-mile-long spacecraft, oriented in line with the jousting course but 46,500 miles to the left of it? The slender space craft is parallel to the course and is facing the oncoming spacecraft's way. But it's moving sideways toward the gate at one-quarter the speed of light. Light that our approaching spacecraft is going to see is going at half the speed of light relative to the course, but really we mean relative to the side-sliding craft that is not moving toward or away from us very much.

If there were no jousting course to use as a reference, the observer might see the side-sliding craft coming toward him from an angle and headed to cross his bow or himself sliding to the right to pass a motionless craft. When our approacher reaches his entry gate, light from the back end of the spacecraft has come abreast of the nose, but the nose is now at the starting gate, and the image of the tail is 46,500 miles to the left. When the observer has reached the midpoint and the light from the nose and the tail has too, this image of the space needle appears to be stretched diagonally, but only to the length of a hypotenuse. (Maybe slightly curved.) When the watcher reaches that gate, the needle's nose is 93,000 miles to his right and the light from the tail, now only a little more distant than the nose, makes for a less diagonal looking spacecraft.

Now that the needle is moving away from the observer, light (that he sees) is leaving it at five-quarters the speed of light, the nose appears to be four-fifths of its actual distance, but that's also because the image is in the process of flipping. It will not only change from double to two thirds of its actual length but the tail will appear to be leading the nose in its side-sliding. So, as his angle of view has been changing, the side-slipper has seemed to be getting shorter and continues to get shorter after he passes it until it looks two thirds its actual length. When the observer is past the long spaceship, it will look like the tail is side-slipping diagonally ahead of the nose. But as he was approaching, the long skinny craft looked like it was going twice its speed sideways, and it appeared to be very diagonally oriented. Going away, light will have to go one and a half times the speed of light to be seen by the observer. The tail will seem to lead the nose by only 15,500 miles, and it will appear to be moving sideways at only two-thirds of its actual speed, one-sixth of the speed of light. And this on top of appearing to be going away from the observer at two-thirds its actual speed or one-third the speed of light. We use the jousting course to describe where the spacecrafts are going but we do our calculations with speeds relative to each other.

Back to the jousting course once more. Let's take two 93,000-mile-long spacecrafts that are going toward each other. One is going at one quarter the speed of light; the other at half the speed of light. Let's call them "Q" for quarter speed and "H" for half speed. Light (that can be seen by the other) leaves each spaceship at one quarter of the speed of light. Light from Q is going at half of the speed of light relative to the course and light from H is going three quarters of the speed of light relative to the course. Let's start with both of them 558,000 miles apart—we can't think in terms of the jousting course because we have to think in terms of combined motion now. We're describing how each sees the other, not a motionless camera. They're coming toward each other at three quarters of the speed of light. Each sees the other coming at three times the speed of light (six times one's actual speed and 12 times the slower's), and each appears to the other to be four times its actual length. After they pass and are going away from each other, they appear to each other to be going three-sevenths of the speed of light and to be four-sevenths of their actual length.

How about what each spacecraft sees on the jousting course? Let's start them at 186,000 miles away from their entry gates, same distance as before. But now they are measuring the other's speed relative to the jousting course and subtracting out their own speed. Q is seeing a jousting course that is four thirds its actual length and H is seeing a jousting course that is twice as long as it really is. After two seconds, when Q is 93,000 miles away from its entry gate, the light from its tail has caught up with its nose. At this time, H has reached its gate and light from its tail has also caught up with its nose. To H, Q appears to be twice as long as it really is, and H appears to be three times as long as it really is. (Remember, they each see the jousting course stretched differently and now they are using the course as a reference for length.) This is why we have to use combined speeds instead of reference points; the references get distorted when the observer is moving too.

What I'm calling relativity here, is image lag, but it is caused by vessels traveling at speeds that cause light to be a little too slow for the images to keep up with what's happening. This optical illusion relativity works if light travels at different speeds, and it's a much better explanation for high speed visual phenomena than light traveling at only one speed and space contracting and time slowing down. But the observer's speed influences this sort of relativity too. His speed affects how he sees the motion and the image just as much as the performer's speed affects it so the combined speed is effectively the real speed.

Chapter Eight
Super Lucent Speed

What would we see if the spacecraft were going faster than the speed of light? The speed of light isn't really a barrier. It only seems to be so. This speed requirement exists for light if it is to be detectable and seemingly also for subatomic particles being accelerated in a cyclotron. (More about that further along.) What if the 93,000-mile-long space needle were traveling toward you at twice the speed of light? When the front end reached the camera, it would seem to pop out of a hole in space beside you. The light that had been traveling away from the rear-showing part of the spacecraft to the rear, leaving it at the speed of light would now hit you from where the spacecraft had been. This would make the spacecraft appear to be traveling in the opposite direction. (Nose first!) You are seeing a reverse image: the light you are seeing was going away from you, but you are seeing it nonetheless because it was actually coming toward you at the correct speed. When the (real, not image) spacecraft's nose entered the gate 93,000 miles from your camera in the middle of the course, it took the light from that nose (which you could hardly see because its image was blocked by the tail) one half of a second to reach your camera. One half of a second after the nose entered the gate, the spacecraft's tail (93,000 miles behind the gate going 372,000 miles per second) passed the midpoint and the camera saw both images, the nose at the entry gate and the tail beside the camera. (You saw the nose pop out of nowhere beside you a quarter second before that.) The spacecraft appears to be its actual length and appears to go back to where it came from at twice the speed of light. (The nose appeared to go its 93,000 miles back to the gate in one quarter of a second.) Using our equations, we see:

$$Va = Vr \times Co / (C - Vr)$$
$$Va = 2 \times -1 / (1 - 2) = 2 \times 1 / 1 = 2$$

Apparent speed is two times the speed of light.

$$La = Lr \times Co / (C - Vr)$$
$$La = 1 \times -1 / (1 - 2) = 1$$

The backward image appears to be as long as it actually is.

At the same time, the present image of the actually going away spacecraft comes out of the same hole in space going the opposite direction of the reverse image. Light leaves the double-light speed spacecraft at three times the speed of light to be going at the speed of light for our other camera mounted in that midpoint satellite, but pointed in the opposite direction toward where the spacecraft is actually going. One-sixth of a second before the (real) spaceship has completely passed our camera in the middle (one-twelfth of a second after the nose left the hole), light that leaves its nose 31,000 miles away from the midpoint camera will depart the nose and after that one-sixth second is up, will cross paths with the spacecraft's tail right next to our camera. The spacecraft's tail covers 62,000 miles from 31,000 miles after the entry gate to the midcourse camera, in the same time, one-sixth of a second, as the light from the nose goes 31,000 miles back to the camera. The spacecraft will appear to be one-third of its actual length and be going one-third of its actual speed. Again, using the equations to check to see if I have this right:

$$Va = 2 \times 1 / (1 + 2) = 2/3$$

The spaceship appears to be going 2/3 of the speed of light or 1/3 of its actual speed.

$$La = 1 \times 1 / (1 + 2) = 1/3$$

The spaceship appears to be one third of its actual length.

What if the spaceship is coming toward us at one-and-one-quarter the speed of light? To use the equation for approaching spaceships (again, we are using terms in multiples of the speed of light and substituting "1" for "Co" since the observer is stationary.):

$$Va = 5/4 \times -1 / (1 - 5/4) = 5/4 \times 4 = 5$$

Apparent speed is five times the speed of light.

La = 93,000 x -1 / (1 - 5/4) = 372,000

The apparent length of the spaceship is four times its real length. (If we did not use the negative value for Co in the equation, then the reason for the numbers for our answers being negative would be to imply that the inverted image and going backward must be represented by negative numbers.)

If the spaceship were going 100 times the speed of light, the false image would appear to be going back the way it came at slightly greater than the speed of light and the real image would look as though it were going away in the other direction at a speed slightly slower than the speed of light. Its image would be one-ninety-ninth its real length coming (going back) and 1/101th of its real length going on its speedy way. In both cases, they approach the speed of light asymptotically but coming toward the line from different directions.

Perhaps light cannot stand still, so it is impossible to see an object coming toward you at the speed of light (until it has reached you and passed you), but an optical illusion would not prevent actual objects from going at the speed of light or faster.

**It must be
that light
comes out of
its sources at a
considerable
range of
speeds.**

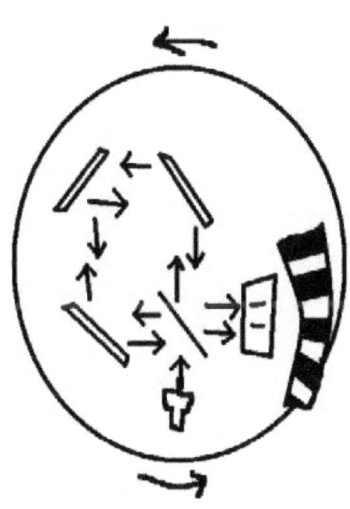

The Sagnac common path (but opposite
directions) interferometer shows that when a
target moves toward an equally receding source
of light the light's path is shortened.

While the apparatus is rotating, in the extremely
brief time between light leaving the half silvered
mirror and it hitting the diffraction slits, the path
has shortened by a significant portion of a wave
length. Since light does not gain momentum from
its source's motion and there is no one universal
frame of reference for the correct speed of light,

Chapter Nine
An Actual Example of Relativity

It has been said that global positioning satellites have to be set to run 38 nanoseconds per second slower than normal clocks. The reason given for this is that up there in orbit they are out of Earth's gravity well. Down here on Earth, gravity is slowing down time. But there is said to be another effect. Their speed going around the world so fast (around four and six tenths miles per second) causes time to slow down for them. Their clocks would have to be slowed down even more to compensate for lower gravity if they weren't being slowed down by going so fast. So, it is because of gravity's effect on time that the satellite's clock is built to go slower to keep up with the rest of the global positioning system. Now on the other hand, the General Theory of Relativity that time goes slower in a stronger gravitational field and a satellite being farther out of Earth's gravity field, would experience time going faster could be an explanation for the results of the Michelson Morley experiment. In the early twentieth century, a Frenchman, Georges Sagnac, performed an experiment similar to the Michelson Morley one. He used a semitransparent mirror to split a beam of light, but instead of sending the two beams back and forth with other mirrors at 90-degree angles to see if Earth's motion had an effect on how far light had to go, he used mirrors to send light around in a circle opposite ways and back to a diffraction grating. Now when he rotated this ring of light, the pattern of light and dark bars did change! So reliably, in fact, that today these things are used in inertial navigation devices in place of gyro compasses. Can it be that massive bodies with large gravitational fields tend to nail down motion, that light moving relative to them is moving relative to stasis? Rotating the Sagnac device will bring faster and slower light to the two slits. Only stretched by a fraction of a wavelength to change the bars image, but more than Michelson Morley's motion of the Earth. Again, we need more and different experiments to see what effect the earth has on light.

Going back to the Special Theory of Relativity, perhaps the satellite's speed does reduce gravity's effect, but relativity is meant to explain how light goes the same speed for two observers moving at high speeds in different directions. The Sagnac Effect comes up again as an issue regarding the problem that when the satellite is traveling south, GPS users and tracking stations are moving away from it to the left, faster at the equator, and to the right when it is going north. Same as in the wheel of light, light is going faster or slower and must be compensated for. The same thing is going on due to the satellite, itself, moving and compensations are made for that too. It is said that one of the principles that scientists observe in calculations for the GPS system is that the speed of light does not change. In a sense, this is true. No light changes its speed. (Except for when it comes away from the gravitational pull of stars, or possibly when it bounces off rapidly receding mirrors and of course when elevating electron orbits in the process of refraction with glass or other objects.) It is just that there is lots of light coming out all kinds of sources and going many different speeds as well as going the one detectable speed of light. The detectable speed is the one observers going the correct speed and direction will see. But if it is going the wrong speed then it is undetectable. The same goes for radio waves. So GPS users' time signals coming from behind and in front of the fast-moving satellite are travelling at the speed of light for the satellite but faster and slower for the user and the Earth on which the distance is being measured. Like the Sagnac effect from Earth's rotation, this too must be compensated for.

Take, for example, a satellite approaching the user and going about four point six miles per second. The user puts out a radio pulse to the satellite that, for the satellite to be able to receive it, is moving slightly slower than 186,172 miles per second. (I gotta be more precise than 186,000 here.) This is the relativity factor that must be corrected for. The computer knows each satellite's vector and compensates for this relativity factor which could amount to a few hundred nanoseconds per second or a few hundred yards if the satellites were about one or more thousand miles away from the user and not being corrected. (There are 24 GPS satellites in use and usually four are above the horizon for any place on Earth.) One of the other satellites is no doubt going away from this user and that same radio beat (that this satellite also receives) goes 4.6 miles per second faster than the ordinary speed of light or radio waves. A third satellite somewhere else gets a commensurately later or earlier signal so they have to come to terms regarding their different motions. The computer has to compensate for this and knowing each satellite's position, speed and direction, it inserts compensations into the calculations. And it includes the correction for the earth's rotation, that affects the speed that the radio waves traveled from the user as well—the Sagnac effect. Whether you choose to attribute the variations of relativity to different speeds of different light, time being slowed, or space contracting, the calculations give correct answers. It is new and different experiments that will tell us what is responsible for the effect.

The particles accelerated in cyclotrons and more advanced accelerators are also to some degree an example of relativity. Is it true that the speed of light is no barrier? If that is so then why can't scientists accelerate protons or other subatomic particles faster than the speed of light in advanced particle accelerators? One answer is that as they approach the speed of light, they gain mass and that the energy that you put into making them go faster just makes them heavier. That is evidenced by the size of their impact on the target that they hit. There is some similarity here to light being undetectable unless it is going 186,000 mps relative to the observer. I also suspect that a spacecraft approaching at the speed of light would have a great deal of light and other electromagnetic energy of a very kinetic nature, that would be very detectable by the observer if he were in close proximity to the arriving vessel. I can't honestly equate this with a gain of mass, but someday someone will perform an experiment of this nature, and we will find out what is accompanying or is affecting that small but not microscopic accelerated object. Going back to the accelerated particle, I also suspect that if suddenly, the target in the cyclotron zoomed away from the approaching proton at a high speed but lesser speed than the particle, that even though the proton did not slow down in the slightest, the magnitude of the impact would not only indicate a loss of kinetic energy but the proton's loss of mass as well. There is also the possibility that trying to accelerate a particle such as a proton to a speed faster than light using electromagnets just doesn't work. No matter how quickly you switch to the next electromagnet ahead of the particle, stationary electromagnets can't accelerate a particle faster than light. Electromotive force may have a top speed.

On the other hand, maybe it is that they have accelerated particles faster than light but then the particles lost mass (interactivity) and the diminished impact led to the wrong conclusion that the speed was less than it actually was.

So called proofs of relativity will face new tests in new venues sooner or later and will be better understood.

Chapter Ten

Why are There No Examples of Things Going Faster than Light?

What about something that is easier to observe than a proton or an electron? In this expanding universe of ours, everything is going outward and away from the center. Out on the edge of the universe things are going very faster relative to the center. Whatever is going at a much different speed at a far distant edge is not coming our way. (Except light that must have departed at a greater speed than 186,000 miles per second to escape the gravitational pull of the star that it came out of and to be going light speed in our rapidly receding galaxy.) So, let's suppose we built a satellite like Sputnik (conveniently small and only beeps a radio pulse), put it on a massive multi-stage rocket, took it to a place two billion miles from the sun—more than 20 times beyond Earth's orbit, and launched it toward the Sun. Our purpose would be to see if we could get its speed up to 200,000 mps, faster than the speed of light. The sun's gravitational pull would help a little at the end of its ride. Its average speed would be 100,000 miles per second so it would have 20,000 seconds to accomplish this speed test. It would have to accelerate 10 miles per second per second or 36,000 miles per hour per second. That's like being shot from a high velocity rifle that was, itself, shot from a bigger such rifle 36 times every second for 20,000 seconds. So, we are stuck with using protons and electrons, subatomic particles so small that they are theoretical and require giant arrays of electro-magnets to propel them, and become detectable only when they hit targets similar to Wilson cloud chambers to analyze the results of the impact. Two hundred years ago, it was well established that supersonic speed was impossible, and doctors marveled that a soldier could be killed by a cannonball that had missed him. Scientists have collided particles going one way with particles going the other way and have declared the results to show that the speed of light is the limit for the particles' combined speeds. Again, in such an experiment, we see the (less) shattering results of the collision that are attributed to the impossibility of speed greater than light but we don't know for sure how the particles behaved before the impact. My suggestion is that it is not added mass but increased interactivity which is at its maximum effect only when particles collide at a combined speed of 186,000 miles per second. This interactivity is not the same as mass colliding with mass. It is only an enhancement of mass that comes along to its highest degree when the mass meets other mass at the correct (detectable) speed of light. Not only if the combined speeds are less but also if they are greater, then the enhancement is diminished. When the two particles collide at a relative speed faster than light, they do so with increased kinetic energy but less added "mass". Maybe that "punch" is just left behind at super lucent speeds. (Or just lost altogether.) As the combined speeds of both particles surpasses the speed of light, the loss of this interactivity with each other balances out the gain in kinetic energy. The energy in the collision does not increase so the conclusion is that the combined speeds can never reach the speed of light.

Chapter Eleven
Another Experiment

So what experiment do I propose to find out whether or not the stuff of light goes at different speeds faster and slower than the speed at which we see light travel? If the twirling mirror/wheel of mirrors experiment just isn't feasible then despite the difficulty in accelerating satellites to significant fractions of the speed of light, I suggest this: Set two satellites on opposite courses at high (as high as possible) speeds going toward and away from a massive planet such as Jupiter. Both will have cameras pointed at the large planet and the stars that appear close to it and at the instant that the satellites pass each other, they each take a picture. The satellite going away will see faster moving light and the satellite going toward will see slower moving light. Whether by gravity or curvature of space, a massive body attracts light passing close to it. (Perhaps not too significantly, but just enough for our purpose.) Slow light passing Jupiter would spend a little more time being attracted by this massive planet than fast moving light and the slower moving light's course would be bent a little bit more. The different directions that the satellites were moving would not have had any effect on the light while it was near Jupiter so any difference in the curvature would have to be the result of different light not different satellites. I think the satellite going toward would see stars behind Jupiter that would be hidden from the satellite going away. This would show that light travels faster and slower than its given speed. I suppose that the slower light from the stars hidden behind Jupiter and around it would be bluer in the photo (or a spectroscopic analysis) that the toward-moving satellite took. The away-moving satellite's photo would show redder light.

I hope some experiments like this are carried out, and we find out whether light travels at different speeds or space and time adjust themselves to control the speed at which an object travels. I suppose that they will be expensive and difficult, but I think that they would be worth doing.

Oh, and the laser jousting—how did that turn out? After the tournament was over and all of the (very) slow motion replays were reviewed and discussed by the sportscasters, it was realized that those powerful beams of laser light had not had any detectable effect on either spacecraft. The four crew members professed dismay, the maintenance teams were puzzled, and the audience was bored and turned to other interests.

Ending #2

Yes, who knows what the ending would be until an experiment like this is done.

At the instant the two spacecraft crossed paths, both were shattered by each other's electromagnetic shock wave. (In space? Yes, who knows what this "added mass" is.) None of the wreckage hit the camera satellite in the middle and it was strangely unaffected by the shock waves. It was thought for a long time that the two vessels had somehow collided despite the camera's evidence that they had only come very close. Both sides hurled shrill accusations that the other spacecraft had released obstacles in front of their spacecraft but there was no evidence of that either. After a while, both sides came to an acceptable conclusion. It was thought that a few atoms and molecules of cabin atmosphere and other traces of residual fuel or exhaust gas increased in mass when they collided with the oncoming spacecraft at the combined speed of light. Although subsequent tests on the other spacecraft showed very little of this material, it was unthinkable that it could be anything other than these tiny bits of matter that could possess so much energy. Only much later did they find out just how much light and the rest of the electromagnetic spectrum piles up with and ahead of an object traveling at or close to the given speed of light and how all of this energy is released instantly in a collision at the correct speed of light.